The role of the Laboratory Animal Sciences in The Science of Reincarnation

Bob Good

I have a delightful little argument to present. But I caution you at the very beginning, you need to be aware that if you accept the idea of "The Science of Reincarnation" you will create it. The question is what are the ramifications created by the acceptance of this science?

So what does the lab animal community have to do with the science of reincarnation, and is there really a science of reincarnation?

An Ohio University study of heart disease in the 1970's was conducted by feeding quite toxic, high cholesterol diets to rabbits in order to block their arteries, duplicating the effect that such a diet has on human arteries. Consistent results began to appear in all the rabbit groups except for one, which strangely displayed 60 percent fewer symptoms. Nothing in the rabbit's physiology could account for the high tolerance to the diet, until it was discovered by accident that the student who was in charge of feeding these particular rabbits liked to fondle and pet them. He would hold each rabbit lovingly for a few minutes before feeding it; astonishingly, this alone seemed to enable the animals to overcome this toxic diet. Repeat experiments, in which one in which one group of rabbits was treated neutrally while the others were loved, caused similar results. The mechanism that causes such immunity is quite unknown-it is baffling to think that evolution has built into the rabbit mind an immune response that needs to be triggered by human cuddling.

In Paris a tiny heart, propped atop of purpose built scaffolding carried on beating. It was kept alive courtesy of a small team of French scientists, who administered the right combination of oxygen and carbon dioxide, part of the type of state of the art surgical technique used for heart transplants.   In this instance there was no donor or recipient; the heart had long been divested of its owner, a prime male Hartley guinea pig, and the scientists were only interested in the organ itself and how it was about to react. They'd applied acetycholine and histamine, two known vasodilators, and then atropine and mepyramine, both antagonists to the others, and finally measured coronary flow, plus such mechanical changes as heart rate.

There were no surprises here. As expected, the histamine and acetycholine produced increased blood flow in the coronary arteries, while the mepyramine and atropine inhibited it. The only unusual aspect of the experiment was that the agents of change weren't actually pharmacological chemicals but low frequency waves of electromagnetic signals of the cells recorded using a purpose designed transducer and a computer equipped with a sound card. It was these signals which take the form of electromagnetic radiation of less than 20 kilohertz, which were applied to the Guinea pig heart, and were responsible for speeding it up, just as the chemicals would themselves.

The signal could effectively take the place of the chemicals, for the signal is the molecule's signature. The scientific team, which had successfully substituted it for the original were quietly aware of the explosive nature of their achievement. Through their efforts the usual theories of molecular signaling and how cells "talk" to each other had been profoundly modified. They were beginning to demonstrate in the laboratory that each molecule in the universe had a unique frequency and the language it used to speak to the world was a resonating wave.

So what is going on here? So what is happening to the science I learned as a kid?

That science is changing and the change is to use the phrasing from a character out of Monty Python, this is not big, it's friggin huge.

What is changing is something called the metaparadigm. Just what is a metaparadigm?

Princeton physicist and philosopher of science Thomas Kuhn, generally acknowledged to be the leading philosopher and historian of science in the twentieth century, coined the term paradigm, by which he meant the philosophical and theoretical framework within which a scientific discipline builds its theories, laws, and generalizations and conducts the experiments which test those theories and formulations. It is in essence the worldview of that discipline, and when a consensus emerges, paradigm is achieved and that discipline becomes, in Kuhn's terms, a science.

In his classic book The Structure of Scientific Revolutions, he explains that those who are drawn to science and who become scientists are members of a special community dedicated to solving certain very restricted and self-defined problems all of which are defined by the accepted world view or paradigm Kuhn defines the power of paradigms in this way, 'universally recognized scientific achievements [in a given field] that for a time provide model problems and solutions to a community of practitioners."

For scientists who are immersed in it, a paradigm is their world view. Its boundaries outline for them both what the universe contains and, equally important, what it does not contain. The paradigm explains how this universe operates. However, Kuhn recognized that paradigms can and should change because eventually they simply fail to explain observed phenomena. Eventually, anomalies accumulate that the paradigm cannot encompass, and these inadequacies force the paradigm into crisis. Kuhn saw this process of change as revolutionary not evolutionary saying, 'Successive transition from one paradigm to another via revolution is the usual developmental pattern of mature science."

But sometimes the paradigm that encompasses all the scientific paradigms, called the metaparadigm itself changes.

This is an epochal event and we are in the midst of just such a change. So let's review them, these metapardigms. There are only three.

There was the genisis metaparadigm, all science was done under the idea that god created heaven and earth from the year zero until the mid 1800's. As late as the seventeenth century most scientists accepted the "scientific" calculation of James Ussher, an Irish bishop who added up all the generations of the Old Testament and determined that the world was created in 4004bc. That's what I mean by science being done under the genesis metaparadigm. The grand material metaparadigm is the one you and I grew up with.

The grand material metaparadigm has at least 4 critical assumptions.

1 The mind is the result of physiological processes governed by bioelectrical postulates. Man is essentially isolated from the world; his mind is isolated in and from his body.

2 Each consciousness is a discrete entity. The brain is a discrete organ and the home of consciousness which is driven by chemicals and DNA.

3 Organic evolution moves to no specific goal but simply flows to Darwinian survivalism.

4 There is only one space time continuum and it provides for only one reality. Nothing travels faster than the speed of light.

But there is a new metapardigm emerging and it will change everything. And by everything I mean everything.

The New metaparadign is called the **Unified metapardigm** and it makes the following critical assumptions.

1 The communication of the world does not occur in the visible world of Newton but rather in the subatomic world of Werner Heisenberg.

2 Cells and DNA communicate through frequencies.

3 The brain perceives and makes its own record of the world in pulsating waves.

4 Consciousness can reach outside the body.

5 There is more than one space time continuum and information can travel faster than the speed of light.

What I am here to tell you is that the lab animal technician is on the front lines of this scientific revolution. That you need to know what to look for so you can monitor your animal groups, how you should design experiments for them, how quantum biology can manifest itself in your experiments in a macro way, and why it's important to you and the companies you work for.

That the applications of this change transcend not just medical, military and business applications but go to a change that will affect the fabric of our entire social construct.

But because we have a new emerging metaparadigm and a new world view does that incorporate a science of reincarnation?

Is there really a new science called the science of reincarnation and why call it that at all and why is it important? How do you do the science of reincarnation? When did we start studying reincarnation as a science and how would we even do that? Is this simply quackery? And why is it important to Lab Animal technicians and to the companies they work for?

I would tell you that we have been doing the science of reincarnation for the last 60 years.

So where do we see this work being done, and who is doing it? What do they call it? Before I published my book, when I did a google search for The Science of Reincarnation there were no google results for the science of reincarnation.

If you googled the Science of Reincarnation when my book came out it was the only site that said The Science of Reincarnation in the 10 listings on the first page of the search. On the next two pages, the top thirty search results on google over 3 pages there was zero, no mention of the science of reincarnation.

I wrote to Wikipedia and said that there ought to be a science called The Science of Reincarnation. They said no, that category is called reincarnation research which is what the University of Virginia calls their work. OK reincarnation research, so what is that?

1 The observed phenomena **the University of Virginia** has catalogued and studied for the last 60 years is more than

2500 cases of children between the ages of 2 and 7 who claim to have had prior lives. When taken to the place they said they lived not only could they identify the person they were and the family they were in but the family members of the prior deceased family accepted the child as being the person who had lived with them. An example is the video on the home page of www.thescienceofreincarnation.com done by fox news of a child who claimed he was a pilot shot down by the Japanese during WWII. When this 4 year old met the 85 year old veterans form that unit they all said yes, he was that man.

This is called an anomaly, observational phenomena that our current understanding of science cannot explain.

We have scientifically shown with abundant evidence is that there are a significant number of children who remember having lived before, that this is a global phenomenon, that we do not understand why this is so, and that our current scientific paradigm cannot explain it.

Sometimes in science we see an effect before we know its cause. An example of this occurred in the early

1700s. French farmers would say to scientists, "See this rock in my field? It fell from the sky." Scientists at the French Academy of Science knew the farmers were lying, because they could look at the sky and see that it had no rocks. It wasn't until the 1800s that we learned that those rocks were meteorites that, indeed, fell from space. What the scientists in the 1700s were seeing was an effect for which they didn't know the cause.

The scientists at the University of Virginia are painstakingly aggregating these cases so they can mine the data, looking for patterns, and trying to understand the effects they are seeing. They are using the best scientific methods at their disposal and are leaving a record for future generations of scientists to examine. Currently they are loading information on 2500 cases looking at 400 variables.

This effect that the scientists at the University of Virginia are so painstakingly cataloging neither proves nor disproves reincarnation. This study, however, is but one tile in a larger mosaic of the Science of Reincarnation. The question now becomes, are we seeing this type of scientific anomaly in other areas? Are there other events or effects that do not fit our current scientific concepts of how reality operates? The answer is that there are several.

2 At University of Nevada Las Vegas they originated the study of Near Death Experiences; it was Raymond Moody who wrote the book describing this phenomenon. Today there is a nonprofit (501C3) that studies just this and has annual meetings. The following is from Stephan Schwartz, who writes the Schwartz report.

"Recent well-conducted studies reveal that approximately 4.2% of the American public has 'reported" a near-death experience (NDE). The population in the United States is a bit more than 311 million. So we're talking about more than 13 million people in the 'reported" NDE population. The NDE population is almost certainly much greater than 13 million because research has also revealed many people do not immediately report experiences.

Experiencers initially keep quiet for fear of being ridiculed or embarrassed. As one experiencer noted, 'I couldn't talk about it, or I would have been committed to an institution."[1]

Cherie Sutherland, a visiting research fellow in the School of Sociology at the University of New South Wales, a near-death experiencer herself, did a study which showed that 'when people tried to discuss the NDE, 50% of the relatives and 25% of friends rejected the NDE, and 30% of nursing staff, 85% of doctors, and 50% of psychiatrists reacted negatively." On the basis of hundreds of reports, it is easy to see that the experience is so powerful, and often so foreign to the experiencer's view of reality, that it takes many years of inner processing to fully come to terms with it.

I think anyone who takes the time to actually study the corpus of NDE research, and who makes decisions on the basis of data, must come away thinking it is time to stop asking if this experience shows the existence of consciousness independent of the physical organism. And to start asking: how does this work? What influences the processes of nonlocal mind? How do nonlocal consciousness and the physical organism interact? Do we see this elsewhere?

3 At the University of Miami Brian Weiss uses a technique called Past Life regression. His seminars and I have been to and met Brian, fill up and he has a cottage industry giving lectures and selling books. His technique is similar to the techniques used by Norman Emerson at the University of Toronto where he proved Clairvoyance; Robert Jahns work at Princeton proving the intention experiments and Russell Targ and Hal Putoff's work at Stanford taking a real world application of the intention experiments into remote viewing. So we see similar affects with past life regression, clairvoyance, and the intention experiments where non local information arrives by thought alone. Let me give you an example, before I do realize that all protocols are producing similar scenarios and that is a huge statistical aberration, or our reality.

Can humans use this non local process or access it?

4 At the University of Toronto "In 1973, J. Norman Emerson, a senior archeologist and professor, and perhaps the most respected one in Canada, endorsed using psychics in archeology. This break from orthodoxy, advanced by a man in the last years of an influential career, was groundbreaking.

"J. Norman Emerson, senior professor of anthropology at the University of Toronto, founding vice president and former president of the Canadian Archeological Association is considered by many the father of Canadian archeology.

"Close to 90 percent of that nation's professionals in the fields of anthropology and archaeology had trained with Emerson at some point in their careers. His former students include other University of Toronto anthropology faculty members, directors of the National Museum of Canada, professors at other universities, and most government archeologists. His impact on Canada's research in those disciplines has been immense, so when in March of 1973 at the annual gathering of Canada's Archeological Establishment, when he stood to speak, some wondered why a scientist would in the last years of an influential career, suddenly dessert 30 years of orthodoxy for something like a psychic. His answer was because it was the truth."

Emerson's experience with the paranormal began in the 1960s, but it wasn't until he met George McMullen that things that would shake the foundation of archeology began to change.

Emerson would give McMullen an artifact, like a clay pipe or the shard of a bowl. Such artifacts are known as lithics. McMullen would turn it over in his hand, and he would tell Emerson how it was made, how it was used, what Indian group used it, where it was found.

As Emerson said when delivering his paper, McMullen's accuracy ran to about 80 percent. But one day, a man by the name of Jack Miller came to Emerson and asked if McMullen could solve a little mystery. Miller brought with him something made of black stone, about two-thirds the size of an average man's palm, and flat on both sides. It was generally agreed between Miller and Emerson that the stone was argillite. Argillite is a shale-like rock found on the Queen Charlotte Islands off the coast of British Columbia; it is about the hardness of soapstone, which makes it a good material for carving. Miller told Emerson that he knew where the stone had been found—at the bottom of a posthole he had excavated at a site near the town of Skidegate on the Queen Charlotte Islands—and the time period in which it had been worked. Later that evening, Miller gave the piece to McMullen. "The stone, McMullen stated, with the certitude possible only to one totally ignorant of the intellectual information on a subject, was carved by a Black man from Port-au-Prince in the Caribbean, from whence he had been brought to Canada as a slave.

"Emerson was appalled, as he later admitted. Here I had just presented a paper on how good George was and how you could use psychic data, and just a short while later, he was saying something that was patently ridiculous."

Here, then, you have a psychic providing information that flies in the face of established orthodoxy and historical information. There were no black men, historically, on Port Charlotte Island in that time period or in that part of Canada in that time period. McMullen wasn't just wrong. He was completely wrong. It turned out, however, that it wasn't McMullen who was wrong. It was the archeologists who were wrong, and that would be proven.

The first thing that happened was that this lithic was given to other psychics. Emerson gave it to a woman who knew nothing of McMullen's response, and "to his astonishment and excitement, she began immediately telling him that it had been carved by a Black man from Africa who had been the victim of a vast sweeping slave trade, and had been brought to the New World. As she went on, she gave many details about incidents McMullen had not mentioned, but she also corroborated many of the features that he had described."

Other psychics began to confirm McMullen's analysis without ever having spoken to McMullen. This confirmation, however, wasn't the proof.

"Almost two years later, a team of physical anthropologists totally unconnected with Emerson went to British Columbia to do blood analysis on the Indians of the area. Their report, when filed, contained what was for them the disturbing observation that in an area where no Blacks were supposed to have been until Modern Times, one tribe showed unmistakable evidence of a Black forbearer. The tribe in question was the one into which Emerson's psychic team had said the escaped African had married."

This was physical DNA corroboration that a black man had lived with that tribe at that time.

Let me make it clear: the DNA records of the Iroquois Indians contain proof a black man lived in that tribe during the period of time in question; it's irrefutable.

To conclude, we have a respected archeologist using a psychic who, when given a lithic, comes up with a reading that is totally outside of what could possibly be accepted. But it turns out that not only is that same information confirmed by other psychics, it's confirmed by an independent, unrelated research team doing blood analysis of the Indian tribes and now we know scientifically he was there, it a fact.

So how did that happen? What stored information are these psychics accessing and how do we get that into a lab to study? Are the examples we are seeing interconnected?

So the question was simple. If we could receive information, could we send it? Emerson gave his paper in the early 1970's. By the mid 1970's we began to have lab results with the Ohio State rabbit study falling in the middle of that decade.

So if we could receive information, that is, if clairvoyant archeology proved that a person's mind is able to reach through time and space, the next question was, could we send information in the same way? How could we test this ability in the lab? Princeton University's engineering anomalies lab (PEAR) gave that answer beginning in 1979.

5 The Intention Experiments began at Princeton University and were lead by Robert Jahn and Brenda Dunn. If you go to their web site and watch the fifteen-minute video, you will hear Brenda Dunn say "This typifies the dramatic results they have gotten, results that have yet to be integrated into the minds of the current generation of scientists."

These experiments involved programming a computer to produce randomly an equal number of zeros and ones every hour. Every hour there would be 50 percent zeros and 50 percent ones. But because it was random the sequesnce could be 1110100000110. The computer was then connected to a screen that would show two different pictures, for example, a tree and a boat.

This is what the researchers' subjects, average people recruited off the street, would see when they interfaced with the computer. People would sit in front of the screen, and Jahn and Dunne would ask them to make one picture appear larger than the other by their intention alone. People could close their eyes and think "tree, tree, tree," or they could talk to the computer out loud. They were not allowed to touch the computer, so the only way they could affect the computer was by their intent—their thoughts.

Here are the results. Virtually everybody could make one picture appear larger than the other by a margin of 52 percent to 48 percent. If a bonded couple, a man and a woman, sat in front of the computer and did the experiment jointly, the researchers found the computer would produce 54 percent of one picture and 46 percent of the other. If two women, however, sat down together to attempt the experiment, they would get the 54/46 result, but sometimes in the wrong direction. There were some interesting differences between genders and gender pairing, too. Remember how couples seemed to exhibit a stronger influence on the machine?

Couples would influence the machines six times as strongly as individuals. Also, if a couple was not in a relationship, they would still have a complimentary effect on each other. Men had a better chance of getting the machine to do what they wanted, but women had a stronger effect on the machine, though not always in the direction they intended.

This was not an isolated study. These experiments started in 1979 and ran until 1994. Other labs running the same type of experiments got similar results.

The numbers of the total data are interesting. It was a twelve-year period, nearly 2.5 million trials. Of all those trials, it turned out that 52 percent of the trials were in the direction that was intended.

By intention alone, participants in the experiment were able to bend the computers, at least a bit, to their will. They had some influence on it a significant percentage of times. These results, by the way, have been submitted to the U.S. National Research Council, and the council has concluded that the trials Jahn and Dunne conducted could not be explained by chance.

So if it's not chance, what is it? How, on a scientific level, can you explain the effect of influencing a computer by simply wishing?

On the Science of Reincarnation website under Government Studies and Video Library, you're going to find a link to the Princeton University PEAR lab; there you can watch Robert Jahn and Brenda Dunne explain what it is they do.

The best explanation offered on how Jahn and Dunne arrived at the results they got is as follows: Subatomic entities can behave either as particles or as waves. A particle is a precise thing with a set location in space. A wave is diffuse and unbounded, and has a region of influence, which can flow through and interfere with other waves. Jahn and Dunne feel that consciousness has a similar duality. Each individual has its own particulate separateness. That is, you are a defined thing in space, but you are also capable of wavelike behavior, which could flow through any barriers or distance to exchange information, and interact with the world.

At certain moments, this wavelike consciousness can get in resonance with, that is, have the same frequency as, other subatomic matter.

What Jahn and Dunne seem to be saying is that you and the computer develop coherence. That is, the wavelike component of your being gets in resonance, and one can influence the other.

The question we should ask is: can we believe the results they got? Answer: absolutely! They used over a quarter of a million subjects over a twenty-five-year period, and after publishing their results, other labs tried to duplicate their experiments and got the same results. So, we are seeing an effect that we don't quite understand.

Dean Radin said in his book, *The Conscious Universe* that

Just as a photon is both a particle and a wave, perhaps consciousness too has complementary states. In ordinary states, the mind is more particlelike and is firmly localized in space and time. This is supported by the ordinary subjective experience of being an isolated, independent creature. But in unusual, nonordinary states of awareness, our minds may be more wavelike and no longer localized in space or time. This is supported by subjective experiences of timelessness, mystical unity, and psi.

As with particle-wave duality, it is not the case that only one or the other description is true, but *both are true at the same time*. The fact that we have trouble thinking in terms of "both" rather than "either-or" says more about the limitations of language than it does about the nature of reality. If our minds have complementary characteristics, then perhaps we can be more particlelike or more wavelike depending on what we wish to be, or what it is suitable to be at the time, or what we are motivated to become."

If we can influence a machine, can we influence another person or a disease? In fact, we try it all the time. You have all heard about the power of prayer; now it's being taken into the lab and dissected.

6 Targ

Intention studies did not start at Princeton University. Princeton University started their studies in 1979. A decade earlier, **Doctor Bernard Grad of McGill University in Montreal** was interested in determining whether psychic healers could actually transmit energy to patients. Grad had slowed down the growth of plants, by soaking their seeds in salty water. Before he soaked the seeds, he had a healer lay hands on one container of salt water, which was used for one batch of seeds, and soaked the other batch in salt water the healer hadn't touched. The batch exposed to the water treated by the healer grew taller than the other batch.

Grad then did a study to see if the process worked in reverse: would negative feelings have a negative effect on the plants? A man who was being treated for psychotic depression laid his hands on the water for one batch of seeds. Those seeds grew less than those that had been soaked in water that had gone untouched by the depressed patient.

Grad analyzed the water with infrared spectroscopy. The water treated by the healer displayed minor shifts in its molecular structure. It also had decreased hydrogen bonding between molecules, which are similar to water, that has been exposed to magnets. All of this has been confirmed by other scientists.

Grad also found that healers could affect results by intent alone. After controlling for a number of factors he showed that mice with skin wounds healed more quickly when healers treated them. He also found that healers reduced the growth of cancerous tumors in laboratory animals.

So, if we can affect plants and rodents simply by intention, the question becomes: can we affect humans as well? The scientific tests that followed led to an amazing conclusion. The two examples I give you now are typical of results other labs have gotten. These are not isolated findings.

In 1988, a doctor named Randolph Byrd ran a randomized, double-blind study to investigate remote prayer. About half of almost 400 patients in the coronary care unit of a hospital were prayed for by a group outside of the hospital. When the experiment commenced, there were no significant statistical differences between the group that was prayed for and the control group, which was not. The prayer went on for ten months, at the end of which, the group that had been prayed for were doing significantly better than the other group: their symptoms were less severe; they required fewer treatments with antibiotics; spent less time on a ventilator; and had fewer cases of pneumonia.

When you pray for something to happen, you intend or wish for it to happen, and whether you're praying to Jesus, Buddha, or Spider Woman, you're still positively affecting the object of your prayers. What was different in this case was that prayer was being taken into the laboratory and tested.

Studies like this were a preamble for Elisabeth Targ's and Fred Sicher's study, which brought together a variety of healers: Buddhists, Christian healers, evangelicals, a Jewish kabbalist, a Sioux shaman, people who worked with crystals and bells, and a practitioner of

Qigong. The only absolute requirement was that each healer must truly believe that their method worked.

All healing was done remotely, and for subjects, they used a group of advanced AIDS patients. The target and control groups were matched as much as possible in symptoms, T-cell counts, degree of the illness, etc.

When the six-month study was over, 40 percent of the control group was dead, but all ten in the group that had been the target of the remote healing were healthier than they had been at the beginning of the experiment.

No matter what type of healing they used, no matter what their view of a higher being, the healers were dramatically contributing to the physical and psychological well-being of their patients. Another study, by the Mid-American Heart Institute, yielded similar results.

In short they had applied to the human subjects' conditions and test that had been done to the animal study at Ohio University nearly 20 years before and

found humans could be affected just as the animals had been.

7 Russell Targ, Hal Putoff and Stanford University

So if Princeton University proved this phenomenon existed it was Stanford who put it to use by developing protocols for remote viewing. Who would believe this worked? Let's start with the US Army, the CIA, the NSA, and the Russians. Remote viewing is a macro example of non local ability we all possess. Russell Targ who headed the SRI program, (Stanford research Institute) said the following, "I do not believe in ESP, I have seen ESP occur in the laboratory on a day to day basis. I encourage all of you to read about this man. "In science it is just as serious an error to ignore real but unpredictable data as it is to accept false data as true"

Targ and Putoff started the SRI program in the fall of 1972. In essence they used sensitive's, people like Ingo Swann, to look remotely at military targets for the US Government. Their results were spectacular. But let's let Russell talk himself

While working for a CIA program at Stanford Research Institute (SRI) in Menlo Park, California, our psychic viewers were able to find a downed Russian bomber in Africa, describe the health of American hostages in Iran, and locate a kidnapped American general in Italy. We also described Soviet weapons factories in Siberia, observed a Chinese atomic bomb test three days before it occurred, and performed countless other amazing tasks.

My firm conclusion from decades of ESP research is that we misapprehend the physical and psychological nature of the interconnected space-time in which we live. Our internalized perception of nature is often obstructed and obscured by mental noise.

I believe in ESP because I have seen psychic miracles day after day in university- and government-sponsored investigations. It is clear to me, without any doubt that many people can learn to look into remote distances and into the future with great accuracy and reliability. This is what we call unobstructed awareness, or more specifically remote viewing. Remote viewing is a psychic ability that involves learning how to quiet your mind and separate the visual images of the psychic signal from the noise of the uncontrolled chatter of the mind. With remote viewing you can describe and experience objects and events that are shielded from ordinary perception by distance or time. To varying degrees, we all have this ability, and I do not believe that it, or any ESP state, has metaphysical origins. I believe it is just a kind of thinking in which we expand our awareness to perceive nonlocally. And it will become less mysterious as more of us become more skillful. Today there are almost a million Google pages devoted to information about "remote viewing." So at least some people are catching on to the idea that this is not difficult to do.

I was cofounder of the above-mentioned ESP research program at SRI. This twenty-million-dollar, twenty-three-year program, launched during the Cold War, was supported by the CIA, NASA, the Defense Intelligence Agency, Army and Air Force intelligence, and many other government agencies. We developed the technique of remote viewing, which enabled a person to accurately describe and experience places and events blocked from ordinary perception.

So what does this have to do with Lab Animals? The process is called Intuitive medical Diagnosis. Christine King a licensed vetinarian who runs Anima Vet based in Winston Salem North Carolina and offers vet services to farms for horses dogs cats and farm animals has the following to say.

Medical intuition is simply the practice of focusing one's intuitive mind (our sixth sense) on finding solutions to medical problems. I have found it to be an invaluable adjunct to the more conventional means of investigating medical or training/behavioural problems in my patients. Not only does it help me get a better handle on the obvious problem, it alerts me to underlying factors that may not be evident in the normal course of discussion with the caregiver and physical examination using my five senses and just my "rational mind." In fact, it has become such an important component of my practice that I use it on every patient and incorporate it into my professional fee.

Medical intuition helps me "listen in" on what the horse is trying to convey through the illness, injury, or training/behavioural problem and pass that message on to the human caregiver. I must emphasise, however, that it is simply another tool to help me evaluate and treat my patients appropriately. While I will conduct consultations solely by telephone for clients who live a great distance away, I much prefer to combine medical intuitive evaluation with physical examination and other accepted methods of diagnosis.

So what we have here is leading physicists proving a process to a point that many agencies of the US government fund it, independent vetinarians use it as part of their process and major drugs companies ignore it.

The platelet study is a good example of this.

To give some context: the sigma effect each of these protocols (which I will get to in a moment) (Scwartz is referring to displine protocols like meditation) has demonstrated is considerably more powerful than that of 81-milligram aspirin regimes that constitute a foundation of hypertensive disease treatment. If you are middle-aged, or older, and particularly if you are a man, you may well be taking an 81-milligram tablet every day-one of more than 40 million Americans doing so. Jessica Utts, Chairman and professor of statistics at University of California, Irvine, decided to explore just exactly what the difference was between the 'aspirin" effect and that achieved in nonlocal research. Her study compared databases from two protocols, remote viewing and Ganzfeld against the aspirin database. Writing in the Journal of Scientific Exploration she said:

In summary, how are the remote viewing and ganzfeld results different from the antiplatelet and vascular disease conclusions?
•The psi experiments produced stronger results than the antiplatelet experiments, in terms of the magnitude of the effect. There is a 36% increase in the probability of a (result) over chance, from 25% to 34%. There is a 25% reduction in the probability of a vascular problem after taking antiplatelets.

•The antiplatelet studies had more opportunity for fraud and experimenter effects than did the psi experiments.

•The antiplatelet studies were at least as likely to be funded and conducted by those with a vested interest in the outcome as were the psi experiments.

•In both cases, the experiments were heterogeneous in terms of experimental methods and characteristics of the participants.

All of this leads to one interesting question: Why are millions of heart attack and stroke patients consuming antiplatelets on a regular basis, while the results of the psi experiments are only marginally known and acknowledged by the scientific community? The answer may have many aspects, but surely it does not lie in the statistical methods.9

Can I digitize a frequency and sell it over the internet that would produce an effect equal to that of the antiplatelet studies? Or the psi studies?

Now what I am suggesting to you is that these types of phenomenon are very natural in that they will show up in your animal populations and their effect will be seen in your experimental results. These phenomena need to be studied and incorporated into the design of your experiments.

Now everything I just mentioned reflects 4 protocols which have emerged. Let's go back to Stephan Schwartz and listen to his comments again.

It is also important, I think, that we see research on NDEs in the larger context of other disciplines in science. In parapsychological research, for instance, there are now four established, well-replicated, protocols used across many laboratories, researchers, and participants that reliably produce results based on the ability to acquire nonlocal information.

These are the four protocols:

Remote viewing. A protocol to acquire nonlocal information, like describing a person, place, or object about which completely physical sensorium could not provide information. Sitting in a room 2,000 miles away you couldn't know the couple was standing under the mimosa tree while, in the distance, surf crashed against a snow-white beach. Nor that their clothes were natural fibers dyed in soft shades of bronze and gray. But nonlocal perception could give you that information.

Ganzfeld. A protocol similar in intent to remote viewing in which an individual in a state of sensory deprivation provides verifiable information about film clips being shown at another location.

Random event generator (REG) influence. The REG protocol actually can be considered two major protocols. The first constitutes studies in labs where an individual intends to affect the performance of a physical system, such as a random number generator. It is with the second protocol, however, that the real implications of nonlocal consciousness as a social force becomes apparent. Psychologist Roger Nelson, formerly with the Princeton Engineering Anomalies Research group, for several years has run what is called the Global Consciousness Project. Nelson describes it this way, 'Subtle interactions link us with each other and the Earth. When human consciousness becomes coherent and synchronized, the behavior of random systems may change. Quantum event based random number generators (RNGs) produce completely unpredictable sequences of zeroes and ones. But when a great event synchronizes the feelings of millions of people, our network of RNGs becomes subtly structured. The probability is less than one in a billion that the effect is due to chance. The evidence suggests an emerging noosphere, or the unifying field of consciousness described by sages in all cultures. Coherent consciousness creates order in the world."

Presentiment. A measurable psychophysical response that occurs before actual stimulation, such as the dilation of a participant's pupils while staring at monitor screen before the pictures appears. Or, it is a change in brain function before a noise is heard.

Please note: the results achieved are based on the sessions being double or triple blind and properly randomized, and that a pre-agreed analysis, including statistical evaluation for variance from chance, be part of the process. That is, we don't need to get bogged down in antiquated arguments about sleight-of-hand, secret cuing, and the like, although this remains a staple of nonlocal consciousness research criticism. This kind of criticism stopped being apposite several decades ago. As far back as the mid-90s, after studying the data from just one of these four protocols, remote viewing, stalwart denier University of Oregon psychology professor Ray Hyman had to grudgingly admit,

... the experiments [being assessed] were free of the methodological weaknesses that plagued the early ... research ... the ... experiments appear to be free of the more obvious and better known flaws that can invalidate the results of parapsychological investigations. We agree that the effect sizes reported ... are too large and consistent to be dismissed as statistical flukes.6

There are now four protocols in parapsychology with one in a billion odds that it is not chance producing the results -- the Higgs Boson was assessed to be real on the basis of 1 in 300 million odds -- and over 13 million Americans have reported NDEs. It is an act of willful ignorance to deny this. The materialist paradigm is crumbling. It isn't going to happen overnight; it isn't going to go without a fight; but it is failing; and it will be gone for most of science within 30 years. That's my prediction of where the trend is heading. Stephan Schwartz

So where are our minds, we are our memories, where are they?

Perhaps if we could find where our memory is stored we could find ourselves.

But where is the experiment showing the raw effect of memory retention? Is the science for that being done and how is memory storage part of the science of

reincarnation and how do lab animal technicians do that work?

.

In 1987, Jacques Benveniste of INSERM, the French Institute of Science, conducted the following experiment. He took a common antibody, a substance that causes allergies, and exposed it to certain white cells in the blood called basophiles. If you are allergic to a bee sting, molecules of bee venom would not be in your body more than a few seconds before they triggered basophiles to degranulate. This is one of the ways your body protects itself from invading poisons. In his experiment, Beneviste prepared a solution that was certain to trigger degranulation. He exposed the white blood cells to the antigen solution, and got the expected response: the white blood cells degranulated. Benveniste then diluted the antigen solution until it was no longer chemically active, but he still got the reaction. He continued to dilute the antigen solution until he had, essentially, nothing but water. Using a standard formula known as Avogadro's constant, he mathematically confirmed that it was impossible for the water to contain a single molecule of the antigen. When he exposed the basophiles to this extremely diluted antigen solution, they still degranulated.

That's not supposed to happen; his result was outside the parameters of known science. He repeated the experiment seventy times, and asked other research teams to repeat it in Israel, Canada, and Italy; they all came up with the same result.

His interpretation of the results was that the water that had contained the antigen retained an electromagnetic memory of the antigen that had been in it. But such a view is outside the bounds of orthodox science. Science could not explain this apparent instance of homoeopathy in Benveniste's research.

Benveniste was criticized for his findings, and others tried to demonstrate that his work was flawed, but the labs replicating his experiment were too tough a nut for the naysayers of the scientific world to crack. This was really the beginning of digital biology.

If Benveniste's interpretation of his experiment was right, there must be other ways to prove it. And if that was possible, what exactly was being proven?

This is a model of memory at a quantum level, the wave signature is being digitized and stored and can be remembered across generations, example the monarch butterfly.

Monarch Butterflies are born in central Mexico and fly north. That generation dies in the middle of the United States; the generation born there flies north to Canada where they die. The next generation flies back to Mexico to the same valley, a place neither they nor their parents have ever been.

When we capture them say in Kansas and fly them on a jet to New York and release them, they start flying on a heading as though they were in Kansas heading to Mexico. But soon they adjust their heading and keep flying to Mexico, right to the valley where they originate from.

Where is that memory stored and how is it stored? DNA self replicates but here we have a memory of a location reaching across generations. Is that the same thing we are seeing manifested in the children at UVA? Or are we accessing a memory out of a cloud, something we see in computer storage. Where is memory stored and how is it accessed? I learned when I went to school that memory is stored in the brain. That information is no longer valid. So where is memory stored?

Karl Lashley is a neuropsychologist who spent many years

investigating where in the body memory is stored. His methodology was

to train rats to do various tasks, like negotiating a maze. After they'd

demonstrated that the task had been learned, he would surgically remove

different parts of their brains. (Crude, and possibly cruel, but this was the

forties, before scientists worried about things like that.) He assumed that

he would eventually find exactly the portion of the brain where the

learned tasks were stored.

Instead, what he found was that no matter what part of the brain

was removed, the memory stayed intact. Even when he took out large parts of a rat's brain, as long as they had motor skills left, they could still drag themselves through the maze, the way they had learned to.

This made no sense if memory had a specific storage location. It seemed that instead of being stored somewhere in the brain, memory was stored *everywhere* in the brain. When a colleague of Lashley's, a younger neurosurgeon named Karl Pribam, read an article in the mid-sixties on the first holograms, he had a flash of insight, and for the first time thought he could put together a theory to explain Lashley's rats.

Holography is a product of *interference*, the crisscrossing pattern that's generated when two waves intersect. For example, if you drop a pebble into a pond, it creates a series of concentric waves that spread outward. If you drop a second pebble in, the second set of waves will expand and the two sets of waves will pass through each other. The pattern this produces is an interference pattern, anything that takes the form of waves, like light waves or radio waves, can generate an interference pattern. Laser light, an extremely coherent form of light, is particularly good at creating interference patterns. So here is how you create a hologram. You take a laser and run the light through a beam splitter, creating two separate beams from the same beam of light. Using mirrors, you bounce one of the beams off the object you wish to make a hologram of, and then direct that beam through a holographic plate.

The second beam of light is bounced off another mirror toward the same holographic plate. The two beams create an interference pattern on the holographic plate, like the pattern those two pebbles make in the pond.

The image on the film looks nothing at all like the original object. It looks like sets of concentric rings. But when another laser beam is directed through the film, a three-dimensional image of the original object appears behind it.

You can walk around a holographic projection and observe it from any angle, just like a physical three-dimensional object. But the thing that is even more amazing, and what got Pribram all worked up is this. If you cut the holographic film in half, and shine a laser through it, each half will project the entire image. If the halves are divided again, an entire image can still be created from each portion of the film. The image will get less and less distinct, grow fuzzier with each division, but the entire image is contained in each piece of the film.

Let me say that again. Unlike normal photographs, every small fragment of a piece of holographic film contains all the information recorded in the whole.

This is why Pribram got so excited, because he finally saw a model that explained how memory was stored. If a fragment of holographic film could contain all the information contained in the original image, why couldn't every part of the brain contain everything to recall a whole memory? It would explain why the rats could remember the maze when they had parts of their brains removed. Portions of their brains retained their whole memories just as portions of the hologram retained the entire image of what had been photographed.

So, do we have evidence of this kind of thing at the macro level?

As reported in *The Guardian*, a woman named Claire Sylvia underwent a heart and double lung transplant in 1988. Afterward, she began to crave foods she had never cared for, found herself attracted to women, and started having dreams about a man named Tim. She hunted down the family of her donor, who turned out to be a young man named Tim, and her cravings were for all his favorite foods.

What does this show? Your memory is stored holographically throughout your entire body. **In short it is stored as a wave pattern.** And since your mind must be able to access your memory, your mind must be able to operate throughout your entire body.

At the Univerity of Iowa

Neuroscientists have believed that three brain regions are critical for self-awareness: the insular cortex, the anterior cingulate cortex, and the medial prefrontal cortex. However, a research team led by the University of Iowa has challenged this theory by showing that self-awareness is more a product of a diffuse patchwork of pathways in the brain - including other regions - rather than confined to specific areas.

The conclusions came from a rare opportunity to study a person with extensive brain damage to the three regions believed critical for self-awareness. The person, a 57-year-old, college-educated man known as "Patient R," passed all standard tests of self-awareness. He also displayed repeated self-recognition, both when looking in the mirror and when identifying himself in unaltered photographs taken during all periods of his life.

"What this research clearly shows is that self-awareness corresponds to a brain process that cannot be localized to a single region of the brain,"

But this still needs a further explanation. If mind and memory are diffused wave patterns, then there has to be a theory of how they interact with our bodies on the macro level.

Apparently, we are not there yet. There is no well-developed, sophisticated theory to explain how this fits in our current understanding of biology. We measure the electromagnetic pulses of the heart in an electrocardiogram, and we're always improving our imaging of brain scans, but the energy patterns of the body itself are only at the early stages of being mapped, and it is these energy pulses that create the same kind of interference patterns that you see in a hologram. These energy pulses also create auras, which we're able to capture on certain types of film, and it is these auras that can reach out and intersect other energy pulses that everything around us emits. We are, in essence, reducing our bodies into simple energy for such analyses.

In the 1600s, blood was nothing more than blood, and when a man was ill, it was because he had some bad blood. The doctors cut him and drained out the bad blood. Today, we know that blood is not merely some red liquid coursing through our bodies that needs to be drained when we have any sort of disease. It's a life-giving solution rife with white cells, red cells, plasma, leucocytes, and all sorts of things that are carried through our body. Today, our understanding of our body's energy patterns is just as primitive as the seventeenth century's ideas about blood. Three hundred years from now we may have a more detailed knowledge of the body's energy patterns, because right now we are finding that these energy patterns can intersect and interact with other energy patterns at the quantum level, resulting in the ability to carry vast amounts of information.

Scientifically, then, how do we separate mind and body for the point of this discussion? If our memories are stored holographically throughout our bodies, how do we separate mind and body?

In study after study, I have shown that the human mind has far more ability and power than previously thought. At this point, you must now choose between two scientific models.

The first model is the classic model of the last few centuries. In this model, the mind is the result of our chemical bodies—in scientific terminology, the mind is a result of bio-electrical postulates. If you are person of faith, your belief in a soul will be based on your particular religion. There can be faith in a soul, but there is no empirical evidence of a soul.

The second model is the one I am advocating—that your soul or consciousness came to inhabit your body by means of the scientific processes I have described. After passing from your body, your soul/consciousness will still be you, with your memories and thoughts from this life intact, and will continue in a discrete form.

So I ask you again: which model is more consistent with the science?

I'm sure you're already familiar with the fact that your body regenerates. Let's reflect for a moment on just how dramatic this process is. Actually, 98 percent of the atoms in your body were not there a year ago.

Your stomach lining is replaced every day. You replace your skin every two weeks; the molecules and atoms in your bones are replaced every year. Even the enamel in your teeth is totally replaced every two years. Your body that existed three years ago is gone. And yet, you remember the taste of chocolate ice cream. Your memories from forty years ago are still there. So which is more permanent, your consciousness or your body? We are not talking about speculation here; body replacement is accepted as scientific fact. The body you had two years ago is no longer there. Every atom has been totally replaced.

What does this have to do with reincarnation? A critical argument is built in small, sure steps. According to Deepak Chopra, one of the best-known medical doctors in the U.S., a respected endocrinologist, and an expert in alternative medicine, my body does not create my mind, but my mind creates my body. In a sense, we reincarnate every year. The only part of us that stays the same are our thoughts, memories, and mentality—these constitute the permanent system that we build upon. As Dr. Chopra says, we are not "physical machines that have somehow learned to think," but instead are "thoughts that have learned to create a physical machine."

Chopra states his case with an analogy involving a magnet. Have you ever seen a diagram of magnetic fields and noticed how those fields curve around the magnet, then out and away at both poles? Chopra points out that these fields are invisible until you do the old grade school trick: rest a piece of paper on the magnet, and sprinkle iron filings on the paper. The filings will trace the curve of the fields, even though the actual fields are still invisible.

So are the fields that compose your mind-body activity. Chopra says, continuing that analogy, that "The iron filings moving around are mind-body activity, automatically aligning with the magnetic field, which is intelligence....And the piece of paper? It is the quantum mechanical body, a thin screen that shows exactly what patterns of intelligence are being manifested at the moment."

Chopra points out that the paper is a necessary part of the experiment. If you ever place a magnet near filings without the paper, you've got a mess—they cling to the magnet and you can never seem to clean all of them off.

Chopra says that the paper represents a gap between the body-mind (the filings) and the magnet itself, the *hidden intelligence*. We see the effect—the iron filings following the magnetic field that's under the paper, moving wherever the magnet does—but we don't see the cause, namely, the magnet. Our hidden intelligence remotely controls us, but because of the small gap between it and our body-mind, we can't see it.

So let's look at the quantum part of quantum biology

So, if everything is pulsating waves, from the chemicals within us to the world around us, we and our consciousnesses reside in an ocean of waves of pulsing energy that goes through us and interacts with us. Have we found scientific evidence for this? Yes, we have. It is called the Zero Point Field.

At any given point, where all matter has been removed, and there is a complete and perfect vacuum, and the temperature is absolute zero, we can still detect energy. In her book, *The Field*, Lynne McTaggert calls this zero point energy, but it could also be dark matter, because matter and energy are basically one and the same.

This energy pops in and out of existence, or at least in and out of our one temporal and three spatial planes.

If they have a consistent existence, it is in a dimensional place we cannot access with our senses.

These particles exist in wave form. They do not become particles until they are observed. If we apply this to human consciousness, it means we existed in wave form before we were born. In Zen Buddhism, there is a question asked of a monk. "Where do you go when you die?" He answers, "Back to the place I was before I was born." I will show how religious beliefs held across the world support the scientific findings we've been discussing, and how these scientific findings support doctrines of the world's religions. But for the moment, let's stay on the science.

Quantum physicists discovered that any quantum particle would collapse into a set entity when it was observed or when a measurement was taken. A quantum particle would react to observation and become a definite object. Remember how water never seems to boil when you're watching? It's as if the water didn't exist until you looked into the pot, at which point it would come into being.

Looking at these particles, observing them and measuring them, force them into a set state. Until we looked at them, these particles could only be considered as probably being at a point in time and space until our observation froze them in a set state.

Now think about how particles act and the nature of reality. It means that you, the observer, bring an object into being by observing it. Nothing exists until we observe it.

So if you were to look around the room, everything exists around you as a wave form. That is its basic component. You are yourself a wave form, and when you look at anything in the room, it comes into being, because you are observing it.

If your wave form signature is limiting its perception to perceive other wave form signatures as materiality, and you view your spirit as being within you, as so many religions do, then what you've just read is an application of the standard scientific explanation for how we see to describe in scientific terms your soul and its perception.

In short, everything around us is in a wave form until we look at it, and then it appears to collapse into a particle/object.

If we can digitize chemicals and they perform the same as either chemicals or digitized oscillating frequencies, then can I digitize you? I know that some readers will immediately think of the transporter on Star Trek. If your mind can reach out through time and space in a form that is wavelike, then can that same mind retain its frame of reference without your body?

Doesn't it do that already when it reaches out through space and time, as we proved in the clairvoyant studies or in the Intention Experiments? Is that proof that your mind exists outside your body without losing its essence?

We have already seen that it can do this. So can it do this after your body dies? Does it need your body at all? What role does your body play? What role does your brain play?

In *The Field*, McTaggert says "Our brain primarily talks to itself and to the rest of the body not with words or images, or even bits or chemical impulses, but in the language of wave interference: the language of phase, amplitude and frequency... To know the world is literally to be on its wavelength."

She goes even further when she explains that we see with waves as well. Because we don't see objects in our brains, but in the world, she says, we must be projecting the images we see back out into the world, to perfectly coincide with the original object we're observing. "...we are transforming the timeless, spaceless world of interference patterns into the concrete and discrete world of space and time... the lens of the eye picks up certain interference patterns and then converts them into three-dimensional images."

So, your brain thinks with waves. It sees with waves, smells with waves, and hears with waves. Perception occurs not in the physical macro world, but on the level of quantum particles. We don't actually see physical objects; we see the quantum information that those objects send to us as waves. "Perceiving the world was a matter of tuning in to the Zero Point Field."

Scientists are trying to figure out how this occurs physically and mechanically.

If we're interacting with these interference patterns or energy patterns, which parts of our bodies pick up this information and turn it into the objective, concrete world that we see? McTaggert suggests the possibility that "the microtubules within the cells of dendrites and neurons might be 'light pipes,' acting as 'waveguides' for photons, sending these waves from cell to cell throughout the brain without any loss of energy."

According to this theory, microtubules and the membranes of dendrites represent the Internet of the body. Every neuron of the brain can log on at the same time and speak to every other neuron simultaneously via quantum processes.

Let's develop this theory. We don't just see images in the back of our brain; we perceive them in three dimensions and can even manipulate the images we project. In real time, we are creating our own world with just our observations. It's possible, then, that consciousness is "a global phenomenon that [occurs] everywhere in the body, and not simply in our brains. Consciousness, at its most basic, [is] coherent light."

This explosive discovery about quantum memory leads to the most outrageous idea of all: short-and long-term memory doesn't reside in our brain at all, but instead are stored in the Zero Point Field.

So, what is your brain, then? It should no longer be considered a storage bank—all your memories reside in the Zero Point Field. Therefore, as some scientists, including Ervin Laszlo, would theorize, the brain isn't like a computer's hard drive anymore—it's more like a modem, receiving and sending information back and forth from external storage. That's an enormous change. And if our memory is in the Zero Point Field, a metaphysical Internet of sorts, what does that say about us as beings? **What do you call the overall science that is targeted on finding scientific answers to metaphysical questions?**

The data from the Intention Experiments demonstrate that the human mind, by will alone, can influence a program on a computer. It can also influence people. Coupled with the discovery by quantum physicists that subatomic particles react to almost any particle around them, that they are simultaneously particles and waves of energy, and that observation alone can influence them, the universe becomes a highly interactive, intimately networked system far beyond what common interpretation suggests. It allows the mind to reach beyond time and space to retrieve information. This separates the mind from your body, your momentary reality.

The case for reincarnation benefits from this redefinition of the mind's reach and ability. If the mind is able to reach out and influence other systems, then it is possible that a human's intellect—the soul—can travel beyond the physical brain and live on after the body is deceased.

What you were seeing was a biological display of a new quantum macro expression, in short a quantum cause for a real world observation. When we look for that in the animal population it affects every social and religious construct that we know, it changes all our suppositions.

Let me give you an example. Physicists have entangled 2 particles that did not exist at the same time.

Timeless. In standard entanglement swapping (*top*), entanglement (blue shading) is transferred to photons 1 and 4 by making a measurement on photons 2 and 3. The new experiment (*bottom*) shows that the scheme still works even if photon 1 is destroyed before photon 4 is created.

Now they're just messing with us. Physicists have long known that quantum mechanics allows for a subtle connection between quantum particles called entanglement, in which measuring one particle can instantly set the otherwise uncertain condition, or "state," of another particle—even if it's light years away. Now, experimenters in Israel have shown that they can entangle two photons that don't even exist at the same time.

"It's really cool," says Jeremy O'Brien, an experimenter at the University of Bristol in the United Kingdom, who was not involved in the work. Such time-separated entanglement is predicted by standard quantum theory, O'Brien says, "but it's certainly not widely appreciated, and I don't know if it's been clearly articulated before."

Entanglement is a kind of order that lurks within the uncertainty of quantum theory. Suppose you have a quantum particle of light, or photon. It can be polarized so that it wriggles either vertically or horizontally. The quantum realm is also hazed over with unavoidable uncertainty, and thanks to such quantum uncertainty, a photon can also be polarized vertically and horizontally at the same time. If you then measure the photon, however, you will find it either horizontally polarized or vertically polarized, as the two-ways-at-once state randomly "collapses" one way or the other.

Entanglement can come in if you have two photons. Each can be put into the uncertain vertical-and-horizontal state. However, the photons can be entangled so that their polarizations are correlated even while they remain undetermined. For example, if you measure the first photon and find it horizontally polarized, you'll know that the other photon has instantaneously collapsed into the vertical state and vice versa—no matter how far away it is. Because the collapse happens instantly, Albert Einstein dubbed the effect "spooky action at a distance." It doesn't violate relativity, though: It's impossible to control the outcome of the measurement of the first photon, so the quantum link can't be used to send a message faster than light.

Now Eli Megidish, Hagai Eisenberg, and colleagues at the Hebrew University of Jerusalem have entangled two photons that don't exist at the same time. They start with a scheme known as entanglement swapping. To begin, researchers zap a special crystal with laser light a couple of times to create two entangled pairs of photons, pair 1 and 2 and pair 3 and 4. At the start, photons 1 and 4 are not tangled. But they can be if physicists play the right trick with 2 and 3.

The key is that a measurement "projects" a particle into a definite state -- just as the measurement of a photon collapses it into either vertical or horizontal polarization. So even though photons 2 and 3 start out unentangled, physicists can set up a "projective measurement" that asks, are the two in one of two distinct entangled states or the other? That measurement entangles the photons, even as it absorbs and destroys them. If the researchers select only the events in which photons 2 and 3 end up in, say, the first entangled state, then the measurement also entangles photons 1 and 4. (See diagram, top.) The effect is a bit like joining two pairs of gears to form a four-gear chain: Enmeshing to inner two gears establishes a link between the outer two.

In recent years, physicists have played with the timing in the scheme. For example, last year a team showed that entanglement swapping still works even if they make the projective measurement after they've already measured the polarizations of photons 1 and 4. Now, Eisenberg and colleagues have shown that photons 1 and 4 don't even have to exist at the same time, as they report in a paper in press at *Physical Review Letters*.

To do that, they first create entangled pair 1 and 2 and measure the polarization of 1 right away. Only after that do they create entangled pair 3 and 4 and perform the key projective measurement. Finally, they measure the polarization of photon 4. And even though photons 1 and 4 never coexist, the measurements show that their polarizations still end up entangled. Eisenberg emphasizes that even though in relativity, time measured differently by observers traveling at different speeds, no observer would ever see the two photons as coexisting.

The experiment shows that it's not strictly logical to think of entanglement as a tangible physical property, Eisenberg says. "There is no moment in time in which the two photons coexist," he says, "so you cannot say that the system is entangled at this or that moment." Yet, the phenomenon definitely exists. Anton Zeilinger, a physicist at the University of Vienna, agrees that the experiment demonstrates just how slippery the concepts of quantum mechanics are. "It's really neat because it shows more or less that quantum events are outside our everyday notions of space and time."

So what's the advance good for? Physicists hope to create quantum networks in which protocols like entanglement swapping are used to create quantum links among distant users and transmit uncrackable (but slower than light) secret communications. The new result suggests that when sharing entangled pairs of photons on such a network, a user wouldn't have to wait to see what happens to the photons sent down the line before manipulating the ones kept behind, Eisenberg says. Zeilinger says the result might have other unexpected uses: "This sort of thing opens up people's minds and suddenly somebody has an idea to use it in quantum computing or something."

I will tell you that we do not understand our own reality and need to explore it scientifically. How I explore this topic cannot be dictated to me or you by people of faith. This exploration needs to be totally analytical.

So why call all this "The Science of Reincarnation?

At the beginning of the space program NASA was trying to figure out what to call the people we sent into space. According to Dr. Gamble they held a brainstorming meeting to come up with names. The suggestions ran from spaceman and space pilot to man in a can. Finally they chose astronaut meaning sailor among the stars, the Russians followed suit with Cosmonaut, meaning sailor in the universe.

What they were doing was grabbing the imagination of the world. In short they were selling space exploration. Outer space, astronauts, the heavens, that imagination was selling the space program and the funding.

Certainly inner space is equally important as outer space. But how do we sell that? Why should we want to?

We are all connected, so how does our journey to inner space sell itself. I'm here to tell you folks that this group of scientists couldn't sell their way out of a paper bag.

On a personal note what is occurring amazes me. For brilliant scientists to say everything is connected and to demonstrate that scientifically, and then not connect with each other and then keep the disciplines separate and not unify it under one science is ironic. Look at how the NASA scientists marketed their science, that marketing produced funding and that funding produced product and that product produced profit.

The science of reincarnation is a mosaic, yet not one of the individual groups associate with the others and it creates a false narrative. The study for NDE's has its own 1031C3, which is a nonprofit foundation but it does not include children who claim prior lives, who do not have their own 1031C3 past life regression or clairvoyance or remote viewing.

The name, The Science of Reincarnation does little more than interconnect in a new way was already there and

we were doing anyway. Alone sometimes but with this new interconnection, now we do it more with others and in numbers there is strength.

That is why it is called collectively The Science of Reincarnation. But would that name be accepted?

I already told you If you googled The Science of Reincarnation 6 months ago of the 10 sites that appear on the first page the only site that said The Science of Reincarnation was mine and there was nothing on after that. Today 3 other sites use the science of reincarnation in their header, the few lines that google allows you on their search pages, so now of the 10 sites on the first page 4 sites including my own now use the term The Science of Reincarnation.

On the first 3 pages or the first 30 search results there are more than half a dozen blog posts about this science so from nothing 7 months ago, to one six months ago to more than 33% of all search results 6 months later and growing. The concept of The Science of Reincarnation as a science is being accepted. The fact that I am here addressing you is evidence of that as well.

The Science was already being done; it was just what it was called.

Near death event, the science of soul mates, reincarnation research, it's all a rose by any other name.

So we see its accepted, but why must it be called the science of reincarnation, why is that imperative?

1-It meets Kuhn's criteria as a science. That's simple enough. It is the emerging metapardigm expressing itself. Have we seen this before, how about when William Harvey discovered blood circulated, hematology was born, or how about when we discovered the earth was round?

2-There is no clearing house or place for the information this science would incorporate. But I ask you are Clinical trial data shared sufficiently today?

Are clinical trial data shared sufficiently today? That answer would be no.

The AllTrials campaign asks for all trials to be registered and their results published. **Ben Goldacre** says we need the evidence to make informed decisions about medicines. **John Castellani** (doi:10.1136/bmj.f1881) says mandatory disclosure could affect patient privacy, stifle discovery, and allow competitors or unscrupulous actors to use the information

When discussing transparency it is important to be clear on what is being requested, as obfuscation is sometimes used to avoid discussing simple fixes. At stake are four levels of information about trials: (1) knowledge that a trial has been conducted, from a clinical trials register; (2) a brief summary of a trial's results, in an academic journal article or regulatory summary; (3) longer details about the trial's methods and results, from a clinical study report where available; (4) individual patient data. The AllTrials campaign calls only for the first three to be published.

The status quo is plainly unsatisfactory. The most current review—with no cherry picking permitted—estimates that around half of all trials for the treatments being used today have gone unpublished; and that trials with positive results are twice as likely to be disseminated.[1] This is a problem for both industry and academic trials.

Although some in industry claim that these problems are in the past, in reality all supposed fixes have failed. In 2005, journal editors passed regulations stating that they would publish only registered trials: the evidence now shows these regulations have been widely ignored.[2] In 2007, US legislation was passed requiring all trials since 2008 to post results on clinicaltrials.gov within a year of completion: the best published evidence shows this law has been ignored by 60-90% of trials.[3] If industry representatives believe these problems have been fixed, they should present published evidence to support their case, with methods and results that are available for public scrutiny.

Even if the latest rules on transparency were to be implemented perfectly—starting from now—they would still do nothing to improve the evidence base for the treatments we use today, because they all cover only trials from the past few years. More than 80% of the medicines prescribed this year were generic, and came on the market more than a decade ago. We need the results of trials on these treatments, which are still available, albeit on paper. It is both practical and reasonable to request that these documents should be simply scanned, and shared.

If clinical trials are shared insufficiently then how is this information to be shared when there is nothing to bind the parts together?

3 It drives the imagination.

4-Reason4 it should be called the Science of Reincarnation is the science demands it. This is clearly the most important discovery in all of mankind's history, and we are careening towards it and it cannot be stopped. Our collective social culture promises this continued consciousness. How can we not study our own consciousness and the implication that our consciousness could survive our own deaths? As scientists is it not our obligation to find the truth?

5-You cannot stop it this science from being done. The information needs to be looked at in its totality, aggregated and judged. As the criticisms of the last century on paranormal activity forced the science to become better until so much was better understood this science will take the same course.

So are you going to tell me what Ian Stevenson and Jim Tucker have been doing for the last 50 years not science? Yet think 50 years ago, Virginia, were the religious god fear people of Virginia prepared to call reincarnation a science? Not the Virginia of the 1960's. But it's time; Jim Tucker needs to call what he does science. So to start with Jim Tucker, Raymond Moody, Brian Weiss, Michael Newton Stephan Schwartz Russell Targ Robert Jahn and Dean Radin, need to call this science the science of reincarnation. They need to recognize it as a clearing house as any science that pertains to the many disciplines of reincarnation.

Raymond Moody established the study of near death experiences. Is that not part of the science of reincarnation? Is it not science?

Does Moody do science? He is only looking at one event, near death experiences. So what science is that a part of?

The descriptions of the experiences that Moody's patient match the descriptions of the people that Brian Weiss and Michel Newton get doing past life regression. It is the same descriptive result that J. Norman Emerson got working with clairvoyants. Is that a coincidence? Are both, rather all of these events different sciences and what is that science called? And why can the clairvoyants that Emerson was working with access similar information that the people who were regressed can. Here you name this science? What do you call it?

Did you know that the NDE people have their own convention and it is separate from past life regression conventions or for that matter remote viewing conventions.

The work at Princeton under Jahn and Dunne called the intention experiments and the work at Stanford University under Putoff and Targ covering the same 25 year period proved the human mind can reach outside the body and access information. We also proved that same ability to look inside another body with humans and effect change. We see that in intuitive diagnoses used on animals and the vets are billing for it. **Tell your employers to make no mistake about the fact that there is an untapped income stream in this science.**

What they proved is your consciousness can exist in some way outside your body. It's been validated by Emerson at the University of Toronto in the clairvoyant studies there and the proof is irrefutable. The proof was contained in the DNA record of the Indians.

Studies released recently show clinical trial data in the science world is not shared sufficiently. I just showed you that. Where there is no structure of an overriding science the trial data is shared even less. That's a crime of omission.

I can give you another reason why this is not called the science of reincarnation yet and it's not pretty and this is a crime of commission.

A poll conducted of Louisiana republicans asking who was responsible for the poor federal response to Katrina produced the following factoid. That 29% surveyed felt Obama was at fault and 28% felt Bush was at fault.

Creationists believe dinosaurs ate people.

I had one deacon who read my book The Science of Reincarnation say that he doesn't believe in reincarnation, he believes in resurrection.

What kind of science is a scientist doing if he studies reincarnation? Come on say it with me, it won't hurt he is doing the SCIENCE OF REINCARNATION. See we are all still here.

Now I want to ask you eastern liberal educated fancy pants what you just did?

Did you just make the radical Islamist and the fundamental republican the same? Whoops. You know that the science of reincarnation has shown, that Japanese soldiers who died in Burma came back as Burmese. The German pilots shot down over England came back brits. That there are sex change cases, in fact statistically younger souls reincarnate in one sex but older souls reincarnate 50-50% in each sex. Dam we just made the LGBT community equal to the Islamists and fundamentalists. In fact if your consciousness exists as a wave form wouldn't you want to try all the permutations?

We should teach The Science of Reincarnation as an accredited college level course. We should also teach The Applications of the Science of Reincarnation as accredited University courses.

Wait what did I just say? I said if we accredit courses like the science of reincarnation and the applications of the science e of reincarnation as university courses here in the USofA then the Russians would eventually study it. They would see that sex is not binary but a spectrum that souls go back and forth between sexes from life to life. That LGBT is merely an expression of what it is to be human and in their next life they may be that which they persecute today. How do you not want to teach that at the college level? Will doing this change Russia today? No. But I don't think it will take the 400 years it tool the Catholic Church to say Galileo was right.

But to do that we would have to call it the science of reincarnation.

But what do you call a science that examines this information?

But if this science is accepted here and studied then it must be studies at the Iranian Atomic energy commission and the Indian Institute of Technology, hell they already believe in reincarnation why not study it. What I just said is that by you accepting the science of reincarnation here you are writing the science curriculum in Iran, India, Saudi Arabia and China. Well if you study it and see the Imam and the peasant are equal, that women are equal to men, that sexuality is not binary but a spectrum, then how badly have you turned their intellectual paradigm on its head by having the courage to call this what it is, the science of reincarnation. There is no one in this room that does not think that their environment is not political. Understand that your acceptance of a science that studies reincarnation is going to have global political effects over the course of the next generation.

I would look you all in the eye and tell you you would be negligent as scientists and human being if you didn't call it the science of reincarnation.

You know this will light up the military, the politicians and the lawyers. How do religious leaders tell their followers that another group is Satan when this science has shown the Imam today may be the Nubian female slave tomorrow?

But let's not get caught up in political arguments. We are people of science, is science not egalitarian? Neil Degrass Tyson was being interviewed and he said the beauty of science is what you believe has nothing to do with the truth. You can't do science without the truth and the truth is that this is the science of reincarnation. You can't do science without the courage to do it right. There will be a lot of blow back calling anything the science of reincarnation but I am here to tell you the opponents to this science will get over it.

Ramifications
So what happens if you buy into this idea of a new science? What are the ramifications? What would happen

if you begin to think about how to do this science of reincarnation? What would happen if other people began to think that it was possible to do the science of reincarnation?

You would begin to break down walls in academia for a start. I control this snippet of DNA and you control that one but to see the wavelengths of how they communicate you would need joint ventures.

Look what it does to the legal profession. I don't like lawyers but they would be beside themselves with joy. Look at all the new legal work, ahh the glorious law suits. Could someone claiming a prior incarnation also claim prior property? Who control the benefits of digital aspirin? Could you prescribe medicine not by going to a pharmacy but by opening a program and absorbing a frequency?

Look at religious politics. The social effects are surprising

The science supports the right's fundamental religious beliefs but supports the left's application of them. Sexual discrimination, protection of women's rights, how do you sexually discriminate when the science shows you change gender and race from life to life? Make reincarnation a science and you support the end of this discrimination because it is what the science shows. If each religion is its denomination with its own quirks, then the science produces a common denomination. It filters down the prejudice and hate, makes them all equal and leaves in place their own history and beliefs.

Look at the social change, religion is reduced to a common denomination instead of separate denominations and power shifts from the priest, rabbi imam, and medicine man to the person. Less need for wars, discrimination, and a realization that science is showing there is a personal accountability for action.

More funding for research, look at all those Christian groups that won't do DNA research, can't oppose that any more or sexually discriminate; the science has shown statistically we change race and sex and culture as we go from life to life.

The political overtones of this are enormous. Let's start with the religious right in this country. Is there a woman in this audience who can't see its effect?

How about the 18 year old US or Al Qaeda grunt, the kid who carries the rifle.

The audience in front of me is pretty smart, but a comparative peer group in Pakistan is to use the words of ex Pakistani President Pervez Musharraf "that part of the world is confused". Only 3% of Pakistanis pay taxes you in this audience are supporting them by paying their taxes as the US government gives Pakistan 1.5-2B dollars per year.

But educate that kid, and religious rights of both countries, because most of the service comes from the American south, religious right states, loose recruits because the narratives supporting and propagating these wars are factually false on both sides.

Send that message and you change the world. Not now but in a generation. We already know Kuhn has said that this generation of scientists has to die off before an idea as radical as this can be accepted. You send this idea through electronic media which must stay open and free. That's where the American town hall has gone even as it erodes in America.

Why should anyone believe this? Most of the world already believes in reincarnation, or resurrection, including many people in this room. Science explains our reality, and our science is indicating what we have always believed, this is not all there is and the search for who we are, this inner space, is more demanding and exciting than the search of outer space ever was.

I will tell you why I believe it should be the science of reincarnation. I was that kid who was in the service, the point of the spear. You see this science does more than break down academic walls, it breaks down cultural walls. It gives the military a direct line of communication about who we are to the enemy combatant. It makes us the same. It gives him and his cadre information they can check with their own sources, for example the Indian Institute of Technology or the Iranian Atomic Energy Commission. Buy in from their scientists into something they already believe that is scientifically factual acts as a counter balance in their society to the fundamentalism we fight globally on a daily basis.

In my think tank days we studied something called the McDonald Theory of War. It posits that no two countries that have McDonalds franchises have ever gone to war against each other. At its base is the fact that if we both have McDonald's franchises we have common cultural, and socio economic interests that are the same, and that does not engender a warlike relationship. Can you see that accepting reincarnation as a science would do the same thing? Can you see I cannot put a McDonald's in Abbotabad and every small Taliban town but I can put this science course on their curriculum, not the workers but the cadre's, the educated elite.

Now the syllabus to the study of this science is being written but it is you here in this room who will write that syllabus, and you can't even stop this if you wanted to. But you can improve it; you can speed it along, you can talk about it with your colleges, read the existing syllabus, read the web sites from the universities I have mentioned, and look in archives at the Schwartz report. Quietly watch for it to express its self in your animal populations, because it will. You can help with its birth, but you can't stop its birth.

It whispers to us, it makes promises to our ancestors as a race. Let it revel itself more. Ask the lawyers how to make money from it. How do you patent a wavelength? Do you need a third control group in your animal populations to see if the wave signature of a pharmacological product can be equally effective?

I can't in the speech explain all the permutations and effects of this simple of reincarnation becoming a science. But if I were to just mention some benefits of accepting this new science they would be in no particular order

So what would the benefits be?

It will drive capital investment, it would create jobs. There would be break through discoveries. How can I be sure? Because we know many discoveries are by accident so we will have new accidents in new areas of research. These discoveries would lead to social change to a more equalitarian world because that is how the science treats us, equally. I don't care if you are a Muslim fanatic or a homosexual transvestite, eat a high cholesterol diet and you will have heart problems.

You are already doing this science anyway.

There's money in it.

It would be a new class of interactive science breaking down walls. The science of reincarnation is the interactive study of a variety of disciplines already being done; this science just provides a new angle to a deeper understanding of who we are.

It has applications in, Military, Business, Medicine and Religion.

Everything I just named has to sell to get funding. The military sells fear, you know about business, religion sells hope, but medicine, and these brilliant scientists can't get out of their own heads to sell anything. They need to sell the acceptance of the change in the metapardigm, and in doing it they need to grab the imagination of the world, that's why it has to be The Science of Reincarnation.

6-So what can you do to help? The first thing you can do is call it the science of reincarnation.

First be aware that anomalous results might have a quantum cause as in the Ohio state study I started with.

You might need a second control group to coax out quantum effects. Can cuddling combined with a drug produce a larger result? For my cholesterol I take a cocktail of pravastatin and zetia, can we combine biochemistry and quantum biology in our treatments?

Compliance: The regulators need to build this science into the protocol so scientists get can grant money for studies.

Scientists need to design experiments testing our wave response ability and how to manage better health through this method.

Vendors need to design better equipment. More sophisticated types of faraday cages to block transmissions or facilitate them.

Tech need to look for anomalies that produced results outside of the design range of the experiment they were doing.

The animal welfare act mandates enrichment in primates' dogs and cats this social interaction is a requirement of law. According to Newton's animals reincarnate as well, it is a feature of life is what the anecdotal reports tell us.

So how do you implement The Science of Reincarnation?

What about your employer, will they want you to do the science of reincarnation? Do they want to add it to your research?

They will do this research because there is profit in it.

To look at how the employee, the LAR tech, is effected by and affects this science let's look at it from the POV of your employer. Your employer is driven by profit.

Remember that team of French scientists removed the heart from a guinea pig, and kept it beating (in doing this, they blurred the line between chemistry and physics). While they initially used chemicals to keep the heart functioning, they were also able to use a transducer and computer to create low-frequency electromagnetic waves, keyed to the frequency of the chemicals, which could speed up the heart in the same way the chemicals would. In essence, the electromagnetic signature of the chemicals could do the same thing as the chemicals themselves. There was no difference between the chemical molecule and its frequency in terms of result.

Consider this analogy. If you had a headache, you could either take an aspirin or listen to the electromagnetic signature of aspirin, and both of them would make your headache go away. Each would be equally effective. In essence, the French scientists discovered that molecules speak to each other in oscillating frequencies. So can we talk about medicine that is synthesized on frequency cards you download for one time use, because we have digitized a biochemical compound and can now sell it over the internet? **Can we make them profitable by selling them as computer cards in drug stores?**

Find that signature we find ourselves. You will make discoveries by just doing the science.

I do not have to tell you about the serendipitous nature of drug discovery.

1   Penicillin was discovered by accident because Fleming let some Petri dishes lying around on a window sill.

2   Roach killer is marketed because in a lab at DuPont there was a lot of powdered boric acid that had fallen off the lab benches, it was harmless to humans but they found a lot of dead roaches in it.

3   Saccharine was discovered when a man who was an inorganic chemist went to turn the page of a book and licked his finger and found it to be sweet and then had to look around his lab bench to see what it was.

4   Your own Claritin started as a stomach drug. When given to patients and asked does your stomach feel better they said no but they breathe again.

You will discover in this new science the next class of drugs and drug delivery systems the next multibillion dollar product. That is why you company is interested in your know this and why it's your job to know it.

Monarch Butterflies

Can we find the mechanism that transfers memory from one generation to another? What else is being transferred and what memories do I carry that I am not aware of?

These are just frequencies.

We already heard how Elizabeth Targ has proven in human test groups in terminally ill cancer patients responded to intenders. Can we do the same in animal groups?

By observing the experiment we ourselves change it. As LAR techs we must remember that. We already know we can change behavior and bio medical makeup of the animal population by our actions.

We all know when you enter a primate room you can change the attitude of alpha males to female lab techs by the female lab tech lip smacking. It reduces male aggression in the primates, what brain changes occur in the primate population when that happens? It was only discovered in the last 20 years that elephants communicated with noise by creating it at such a low frequency that we cannot hear it.
These frequencies manifest themselves in ways we cannot explain, with a scientific reason that is beyond our current understanding to explain.

Look at the rabbits from Ohio State.

7 Bring intuitive healer and intuitive diagnosticians in to test on animal population next to your regular tests

If I can change brain wave patterns in a primate population by lip smacking, and I can effect a Hartley guinea pig heart by shooting a specific bandwidth of radiation at it can this research show me a bandwidth that I can shoot at an enemy to change their brain wave patterns, We already shoot pink Floyd at them it by blasting heavy metal music at the enemy so they can't rest. Is there something more subtle and nefarious and how do I defend our troops against it?

Think for a second about the applications of a pulse of electromagnetic radiation, at an enemy army, or one politician. The President of Turkey has already said he is under attack by telekeisis.

What do you test for in animal populations and how do you design experiments?

We know that animals show feeling of loss and can ask for help. In the last 2 weeks on the internet I saw a video of a mother dog actually burying her dead puppy with the nose, covering the body of the pup with loose dust, and a dolphin caught in fishing line on one of its flippers repeatedly swimming up to divers and showing the fin, the divers cut the line off the fin as the dolphin hung motionless in the water as they did it.

Can wave studies be done to digitize the communication signal so we can actually "talk" with another species?

You are on the cutting edge

You are on the cutting edge of this scientific revolution and how you treat your animal populations and how you design experiments with them are crucial to what we will learn from your efforts. Talk about it; educate your peers because the science of reincarnation covers many disciplines. Clairvoyance, why do animals know about tidal waves and earthquakes before humans?

Make no mistake your employer is interested in what I am telling you here today. What does the science of reincarnation have to do with the lab animal technician? Well it doesn't matter if you call it quantum biology or whether you recognize that quantum biology is one discipline in the science of reincarnation, you the lab animal technician will have to know how that science manifests itself in a macro environment.

Do more than report unusual activity, suggest a type of experiment. Any of it could lead to a new class of drugs, or drug delivery system. While you are at it you may just change the world.

The end

www.ingramcontent.com/pod-product-compliance
Lightning Source LLC
Chambersburg PA
CBHW071251170526
45165CB00003B/1296